Multiple Practice Sheet

700 Multiplications

Two digit & Three digit by Single digit

917 x 7 = 6419

88 x 8 = 704

14 x 4 = 56

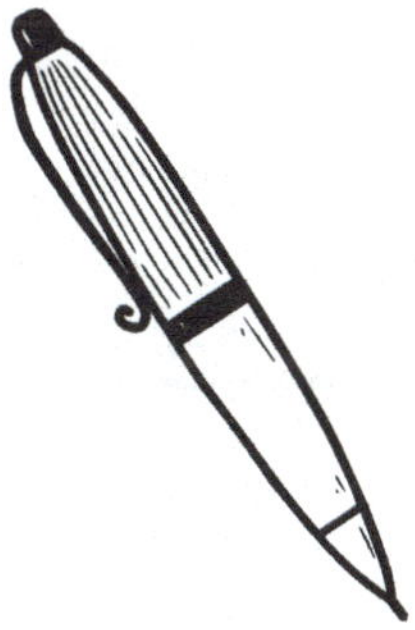

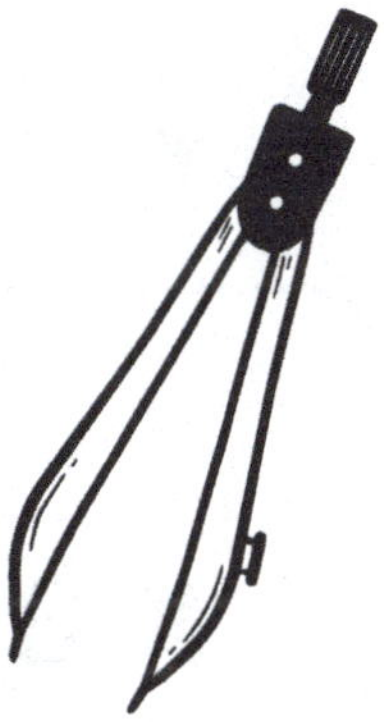

Name: ______________________ Date: __________

Multiplying in Raws

Directions: Solve the Multiplications

60 X 8 = ______

87 X 7 = ______

90 X 9 = ______

85 X 9 = ______

65 X 7 = ______

66 X 8 = ______

36 X 9 = ______

70 X 8 = ______

99 X 3 = ______

60 X 7 = ______

64 X 8 = ______

65 X 2 = ______

98 X 7 = ______

30 X 8 = ______

Name: ______________________ Date: __________

Multiplying in Raws

Directions: Solve the Multiplications

85 X 5 = ______

58 X 7 = ______

35 X 9 = ______

65 X 8 = ______

73 X 6 = ______

90 X 9 = ______

97 X 9 = ______

67 X 5 = ______

78 X 2 = ______

60 X 7 = ______

80 X 8 = ______

78 X 8 = ______

68 X 2 = ______

65 X 8 = ______

Name: ______________________ Date: __________

Multiplying in Raws

Directions: Solve the Multiplications

63 X 8 = ______

57 X 7 = ______

40 X 9 = ______

96 X 9 = ______

99 X 5 = ______

94 X 2 = ______

36 X 9 = ______

65 X 3 = ______

88 X 8 = ______

85 X 7 = ______

50 X 3 = ______

60 X 4 = ______

70 X 8 = ______

69 X 4 = ______

Name: ______________________ Date: __________

Multiplying in Raws

Directions: Solve the Multiplications

66 X 8 = ________

87 X 6 = ________

94 X 9 = ________

32 X 5 = ________

39 X 9 = ________

93 X 2 = ________

35 X 3 = ________

30 X 8 = ________

69 X 3 = ________

72 X 3 = ________

68 X 3 = ________

50 X 8 = ________

88 X 8 = ________

99 X 6 = ________

Name: ________________________ Date: __________

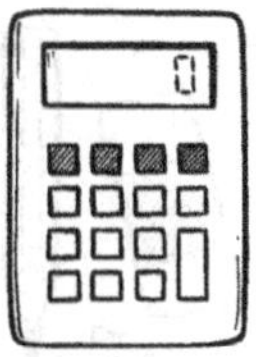

Multiplying in Raws

Directions: Solve the Multiplications

96 X 5 = ____________

77 X 7 = ____________

56 X 9 = ____________

52 X 3 = ____________

94 X 4 = ____________

40 X 9 = ____________

15 X 6 = ____________

84 X 8 = ____________

62 X 7 = ____________

48 X 3 = ____________

89 X 8 = ____________

62 X 7 = ____________

23 X 6 = ____________

65 X 9 = ____________

Name: ______________________ Date: ____________

Multiplying in Raws

Directions: Solve the Multiplications

13 X 8 = ____________

56 X 7 = ____________

87 X 9 = ____________

95 X 5 = ____________

54 X 9 = ____________

76 X 3 = ____________

49 X 9 = ____________

86 X 8 = ____________

32 X 6 = ____________

32 X 9 = ____________

49 X 7 = ____________

56 X 8 = ____________

13 X 3 = ____________

68 X 8 = ____________

Name: ______________________ Date: __________

Multiplying in Raws

Directions: Solve the Multiplications

60 X 2 = ______

82 X 7 = ______

13 X9 = ______

60 X 6 = ______

45 X 9 = ______

36 X 3 = ______

96 X 8 = ______

56 X 6 = ______

78 X 8 = ______

32 X 7 = ______

89 X 5 = ______

86 X 7 = ______

46 X 8 = ______

65 X 3 = ______

Name: ______________________ Date: __________

Multiplying in Raws

Directions: Solve the Multiplications

60 X 8 = ______

87 X 7 = ______

45 X 9 = ______

89 X 9 = ______

35 X 9 = ______

59 X 9 = ______

13 X9 = ______

78 X 6 = ______

31 X 8 = ______

12 X 3 = ______

34 X 4 = ______

48 X 3 = ______

66 X 4 = ______

48 X8 = ______

Name: ______________________ Date: __________

Multiplying in Raws

Directions: Solve the Multiplications

65 X 8 = ______

87 X 3 = ______

56 X 9 = ______

17 X 3 = ______

94 X 9 = ______

79 X 7 = ______

12 X 6 = ______

60 X 4 = ______

95 X 6 = ______

75 X 3 = ______

15 X 9 = ______

61 X 7 = ______

56 X 4 = ______

42 X 7 = ______

Name: ______________________ Date: __________

10

Multiplying in Raws

Directions: Solve the Multiplications

70 X 8 = ________

78 X 7 = ________

32 X 8 = ________

19 X9 = ________

80 X 6 = ________

50 X 4 = ________

70 X 5 = ________

60 X 3 = ________

90 X 4 = ________

63 X 5 = ________

76 X 6 = ________

629 X 7 = ________

63 X8 = ________

83 X 9 = ________

Name: ______________________ Date: __________

Multiplying in Raws

Directions: Solve the Multiplications

604 X 8 = ____________

874 X 7 = ____________

906 X 9 = ____________

806 X 3 = ____________

365 X 4 = ____________

635 X 6 = ____________

325 X 9 = ____________

160 X 7 = ____________

965 X 6 = ____________

341 X 4 = ____________

321 X 2 = ____________

321 X 8 = ____________

874 X 4 = ____________

260 X 9 = ____________

Name: ______________________________ Date: ____________

Multiplying in Raws

Directions: Solve the Multiplications

632 X 8 = ____________

865 X 7 = ____________

954 X 9 = ____________

632 X 7 = ____________

852 X 3 = ____________

741 X 4 = ____________

852 X 7 = ____________

963 X 4 = ____________

369 X 6 = ____________

258 X 8 = ____________

852 X 9 = ____________

147 X 3 = ____________

741 X 2 = ____________

123 X 1 = ____________

Name: ______________________________ Date: __________

Multiplying in Raws

Directions: Solve the Multiplications

632 X 8 = ________

324 X 7 = ________

987 X 9 = ________

789 X 4 = ________

654 X 9 = ________

46 X 3 = ________

321 X 2 = ________

123 X 4 = ________

357 X 6 = ________

753 X 7 = ________

241 X 9 = ________

421 X 3 = ________

689 X 3 = ________

986 X 4 = ________

Name: ______________________ Date: __________

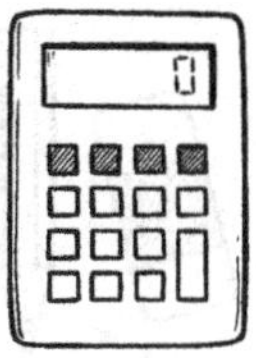

Multiplying in Raws

Directions: Solve the Multiplications

160 X 8 = ________

187 X 7 = ________

930 X 6 = ________

190 X 8 = ________

290 X 9 = ________

390 X 7 = ________

901 X 4 = ________

601 X 7 = ________

609 X 6 = ________

197 X 5 = ________

660 X 8 = ________

670 X 3 = ________

360 X 8 = ________

960 X 3 = ________

Name: ____________________ Date: __________

Multiplying in Raws

Directions: Solve the Multiplications

655 X 8 = ______

834 X 7 = ______

937 X 9 = ______

347 X 4 = ______

157 X 6 = ______

978 X 7 = ______

670 X 9 = ______

662 X 9 = ______

675 X 7 = ______

874 X 5 = ______

324 X 3 = ______

478 X 1 = ______

324 X 9 = ______

357 X 4 = ______

Name: ______________________ Date: __________

Multiplying in Raws

Directions: Solve the Multiplications

160 X 4 = ____________

835 X 7 = ____________

998 X 4 = ____________

955 X 9 = ____________

354 X 7 = ____________

741 X 3 = ____________

475 X 9 = ____________

632 X 8 = ____________

324 X 7 = ____________

345 X 8 = ____________

852 X 3 = ____________

320 X 4 = ____________

120 X 2 = ____________

400 X 8 = ____________

Name: ______________________ Date: __________

Multiplying in Raws

Directions: Solve the Multiplications

460 X 8 = ____________

870 X 7 = ____________

900 X 6 = ____________

907 X 4 = ____________

905 X 9 = ____________

904 X 6 = ____________

903 X 9 = ____________

60 X 7 = ____________

440 X 8 = ____________

604 X 9 = ____________

605 X 7 = ____________

607 X 6 = ____________

609 X 5 = ____________

608 X 4 = ____________

Name: ____________________ Date: __________

Multiplying in Raws

Directions: Solve the Multiplications

645 X 8 = ________

745 X 7 = ________

945 X 5 = ________

907 X 4 = ________

940 X 7 = ________

908 X 9 = ________

964 X 6 = ________

605 X 5 = ________

650 X 8 = ________

604 X 4 = ________

660 X 4 = ________

606 X 9 = ________

60 9 X 6 = ________

960 X 5 = ________

Name: ______________________ Date: __________

Multiplying in Raws

Directions: Solve the Multiplications

640 X 4 = ______

874 X 3 = ______

905 X 9 = ______

940 X 7 = ______

490 X 8 = ______

690 X 7 = ______

966 X 6 = ______

699 X 4 = ______

542 X 3 = ______

789 X 9 = ______

657 X 6 = ______

325 X 7 = ______

624 X 1 = ______

723 X 7 = ______

Name: ______________________ Date: __________20

Multiplying in Raws

Directions: Solve the Multiplications

524 X 3 = ________

324 X 4 = ________

354 X 5 = ________

852 X 9 = ________

754 X 7 = ________

957 X 8 = ________

895 X 4 = ________

507 X 6 = ________

602 X 8 = ________

666 X 8 = ________

555 X 7 = ________

688 X 8 = ________

428 X 9 = ________

625 X 4 = ________

Name: ______________________ Date: ____________

Multiplying in Raws

Directions: Solve the Multiplications

602 X 8= ____________

870 X 7= ____________

904 X 7 = ____________

940 X 6 = ____________

999 X 8 = ____________

690 X 5 = ____________

490 X 9 = ____________

460 X 8 = ____________

640 X 4 = ____________

605 X 6 = ____________

470 X 8 = ____________

604 X 78= ____________

350 X 5 = ____________

352 X 8 = ____________

Name: ______________________ Date: __________

Multiplying in Raws

Directions: Solve the Multiplications

622 X 8 = ______

412 X 4 = ______

487 X 7 = ______

644 X 2 = ______

241 X 3 = ______

365 X 6 = ______

123 X 9 = ______

321 X 8 = ______

321 X 2 = ______

654 X 9 = ______

742 X 1 = ______

321 X 4 = ______

324 X 4 = ______

560 X 6 = ______

Name: ______________________________ Date: __________

Multiplying in Raws

Directions: Solve the Multiplications

222 X 8 = ____________

833 X 2 = ____________

753 X 4 = ____________

415 X 3 = ____________

652 X 5 = ____________

872 X 2 = ____________

358 X 4 = ____________

325 X 4 = ____________

685 X 4 = ____________

389 X 2 = ____________

952 X 2 = ____________

620 X 4 = ____________

333 X 4 = ____________

678 X 6 = ____________

Name: ____________________ Date: ____________

Multiplying in Raws

Directions: Solve the Multiplications

322 X 8 = ______

332 X 7 = ______

666 X 9 = ______

878 X 9 = ______

989 X 9 = ______

323 X 9 = ______

454 X 9 = ______

545 X 8 = ______

656 X 8 = ______

678 X 8 = ______

878 X 8 = ______

654 X 8 = ______

678 X 8 = ______

603 X 8 = ______

Name: ____________________ Date: ____________

Multiplying in Raws

Directions: Solve the Multiplications

635 X 8 = ____________

854 X 7 = ____________

546 X 9 = ____________

656 X 9 = ____________

780 X 9 = ____________

879 X 9 = ____________

935 X 9 = ____________

451 X 8 = ____________

612 X 8 = ____________

678 X 8 = ____________

541 X 8 = ____________

645 X 8 = ____________

785 X 8 = ____________

784 X 8 = ____________

Name: ____________________ Date: __________

Multiplying in Raws

Directions: Solve the Multiplications

124 X 8 = ______

325 X 7 = ______

142 X 2 = ______

921 X 9 = ______

121 X 4 = ______

541 X 7 = ______

213 X 5 = ______

410 X 3 = ______

147 X 4 = ______

159 X 6 = ______

241 X 8 = ______

231 X 7 = ______

312 X 4 = ______

412 X 9 = ______

Name: ______________________ Date: __________

Multiplying in Raws

Directions: Solve the Multiplications

325 X 4 = ____________

210 X 6 = ____________

365 X 4 = ____________

782 X 7 = ____________

310 X 1 = ____________

120 X 7 = ____________

310 X 8 = ____________

412 X 7 = ____________

321 X 8 = ____________

410 X 9 = ____________

321 X 3 = ____________

632 X 4 = ____________

310 X 5 = ____________

103 X 6 = ____________

Name: ______________________ Date: __________

Multiplying in Raws

Directions: Solve the Multiplications

363 X 2 = ____________

333 X 6 = ____________

444 X 3 = ____________

454 X 4 = ____________

654 X 7 = ____________

231 X 5 = ____________

621 X 9 = ____________

632 X 8 = ____________

665 X 4 = ____________

641 X 4 = ____________

324 X 6 = ____________

753 X 4 = ____________

635 X 7 = ____________

852 X 8 = ____________

Name: ____________________ Date: __________

Multiplying in Raws

Directions: Solve the Multiplications

258 X 8 = ____________________

369 X 7 = ____________________

365 X 9 = ____________________

357 X 4 = ____________________

753 X 6 = ____________________

951 X 5 = ____________________

158 X 3 = ____________________

336 X 8 = ____________________

584 X 9 = ____________________

678 X 4 = ____________________

324 X 8 = ____________________

740 X 6 = ____________________

625 X 3 = ____________________

471 X 7 = ____________________

Name: ____________________ Date: __________

Multiplying in Raws

Directions: Solve the Multiplications

632 X 8 = ____________

545 X 7 = ____________

941 X 9 = ____________

321 X 4 = ____________

784 X 4 = ____________

197 X 3 = ____________

379 X 7 = ____________

147 X 7 = ____________

247 X 5 = ____________

374 X 4 = ____________

314 X 8 = ____________

413 X 6 = ____________

521 X 9 = ____________

332 X 5 = ____________

Name: ____________________ Date: __________

Multiplying in Raws

Directions: Solve the Multiplications

475 X 5 = ______

214 X 6 = ______

923 X 7 = ______

143 X 3 = ______

741 X 7 = ______

149 X 3 = ______

324 X 9 = ______

425 X 7 = ______

354 X 8 = ______

123 X 4 = ______

234 X 6 = ______

241 X 5 = ______

324 X 3 = ______

324 X 8 = ______

Name: ______________________ Date: ____________

Multiplying in Raws

Directions: Solve the Multiplications

241 X 7 = ____________

147 X 7 = ____________

598 X 9 = ____________

398 X 4 = ____________

385 X 3 = ____________

347 X 2 = ____________

541 X 1 = ____________

341 X 3 = ____________

341 X 4 = ____________

324 X 8 = ____________

314 X 4 = ____________

214 X 9 = ____________

412 X 7 = ____________

214 X 6 = ____________

Name: ______________________ Date: __________

Multiplying in Raws

Directions: Solve the Multiplications

632 X 8 = ____________

870 X 7 = ____________

930 X 1 = ____________

901 X 6 = ____________

321 X 4 = ____________

214 X 4 = ____________

321 X 6 = ____________

604 X 5 = ____________

641 X 3 = ____________

147 X 8 = ____________

324 X 5 = ____________

147 X 8 = ____________

147 X 6 = ____________

247 X 4 = ____________

Name: ______________________ Date: ____________

Multiplying in Raws

Directions: Solve the Multiplications

148 X 3 = ____________

741 X 7 = ____________

564 X 2 = ____________

247 X 4 = ____________

324 X 9 = ____________

214 X 3 = ____________

327 X 5 = ____________

214 X 9 = ____________

325 X 7 = ____________

574 X 4 = ____________

324 X 8 = ____________

329 X 2 = ____________

951 X 1 = ____________

217 X 7 = ____________

Name: ______________________ Date: __________

Multiplying in Raws

Directions: Solve the Multiplications

247 X 9 = ______

231 X 7 = ______

654 X 2 = ______

456 X 3 = ______

358 X 3 = ______

184 X 9 = ______

247 X 7 = ______

284 X 6 = ______

398 X 8 = ______

574 X 7 = ______

147 X 6 = ______

214 X 3 = ______

320 X 7 = ______

260 X 8 = ______

Name: ____________________ Date: __________

Multiplying in Raws

Directions: Solve the Multiplications

332 X 8 = ______

854 X 4 = ______

324 X 9 = ______

369 X 2 = ______

150 X 3 = ______

724 X 8 = ______

231 X 7 = ______

324 X 6 = ______

624 X 3 = ______

457 X 7 = ______

325 X 4 = ______

124 X 2 = ______

214 X 7 = ______

124 X 8 = ______

Name: ______________________ Date: __________

Multiplying in Raws

Directions: Solve the Multiplications

247 X 4 = ____________

321 X 7 = ____________

254 X 4 = ____________

369 X 9 = ____________

914 X 4 = ____________

321 X 6 = ____________

358 X 7 = ____________

258 X 8 = ____________

247 X 9 = ____________

241 X 7 = ____________

147 X 8 = ____________

214 X 6 = ____________

147 X 8 = ____________

324 X 7 = ____________

Name: ____________________ Date: __________

Multiplying in Raws

Directions: Solve the Multiplications

584 X 5 = ____________________

325 X 7 = ____________________

147 X 4 = ____________________

258 X 4 = ____________________

214 X 7 = ____________________

179 X 8 = ____________________

721 X 9 = ____________________

187 X 7 = ____________________

641 X 8 = ____________________

274 X 9 = ____________________

198 X 7 = ____________________

384 X 8 = ____________________

444 X 9 = ____________________

147 X 3 = ____________________

Name: ____________________ Date: __________

Multiplying in Raws

Directions: Solve the Multiplications

641 X 8 = ______

120 X 7 = ______

324 X 4 = ______

147 X 9 = ______

258 X 2 = ______

247 X 3 = ______

932 X 8 = ______

471 X 5 = ______

666 X 6 = ______

250 X 7 = ______

214 X 4 = ______

324 X 2 = ______

147 X 3 = ______

324 X 5 = ______

Name: ____________________ Date: __________

Multiplying in Raws

Directions: Solve the Multiplications

741 X 7 = ____________

145 X 9 = ____________

241 X 6 = ____________

369 X 7 = ____________

147 X 5 = ____________

347 X 8 = ____________

234 X 7 = ____________

257 X 7 = ____________

247 X 4 = ____________

147 X 2 = ____________

324 X 5 = ____________

741 X 8 = ____________

235 X 6 = ____________

124 X 4 = ____________

Name: ______________________ Date: __________

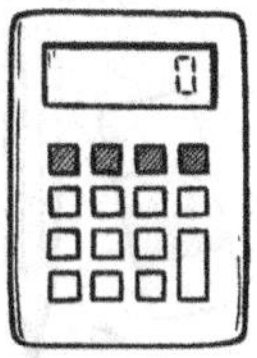

Multiplying in Raws

Directions: Solve the Multiplications

123 X 7 = ____________

124 X 4 = ____________

124 X 3 = ____________

912 X 2 = ____________

124 X 9 = ____________

324 X 7 = ____________

147 X 4 = ____________

147 X 8 = ____________

369 X 6 = ____________

147 X 7 = ____________

478 X 5 = ____________

124 X 9 = ____________

147 X 8 = ____________

475 X 4 = ____________

Name: ______________________ Date: __________

Multiplying in Raws

Directions: Solve the Multiplications

332 X 4 = ______

325 X 6 = ______

541 X 9 = ______

241 X 8 = ______

327 X 7 = ______

147 X 5 = ______

954 X 9 = ______

471 X 7 = ______

147 X 8 = ______

147 X 4 = ______

150 X 6 = ______

328 X 8 = ______

147 X 9 = ______

149 X 6 = ______

Name: ______________________ Date: __________

Multiplying in Raws

Directions: Solve the Multiplications

324 X 8 = ______

257 X 7 = ______

147 X 4 = ______

324 X 6 = ______

247 X 9 = ______

954 X 7 = ______

894 X 3 = ______

364 X 7 = ______

968 X 8 = ______

784 X 7 = ______

324 X 9 = ______

784 X 6 = ______

214 X 4 = ______

320 X 3 = ______

Name: ______________________ Date: __________

Multiplying in Raws

Directions: Solve the Multiplications

623 X 8 = ____________

870 X 7 = ____________

901 X 9 = ____________

999 X 7 = ____________

331 X 6 = ____________

147 X 5 = ____________

324 X 3 = ____________

147 X 3 = ____________

789 X 8 = ____________

324 X 7 = ____________

985 X 3 = ____________

394 X 7 = ____________

459 X 8 = ____________

395 X 3 = ____________

Name: ______________________ Date: ____________

Multiplying in Raws

Directions: Solve the Multiplications

758 X 8 = ______

954 X 3 = ______

347 X 7 = ______

366 X 9 = ______

247 X 7 = ______

124 X 2 = ______

369 X 8 = ______

827 X 8 = ______

365 X 7 = ______

369 X 4 = ______

741 X 2 = ______

987 X 8 = ______

475 X 4 = ______

854 X 8 = ______

Name: ________________________ Date: __________

Multiplying in Raws

Directions: Solve the Multiplications

374 X 3 =

942 X 2 =

248 X 2 =

741 X 3 =

240 X 9 =

254 X 7 =

371 X 8 =

124 X 8 =

731 X 6 =

412 X 6 =

914 X 4 =

214 X 8 =

419 X 9 =

924 X 2 =

Name: ______________________________ Date: ____________

Multiplying in Raws

Directions: Solve the Multiplications

324 X 8 = ______________

214 X 9 = ______________

124 X 4 = ______________

987 X 2 = ______________

214 X 1 = ______________

941 X 6 = ______________

341 X 3 = ______________

241 X 8 = ______________

365 X 7 = ______________

941 X 8 = ______________

241 X 9 = ______________

874 X 3 = ______________

954 X 2 = ______________

632 X 8 = ______________

Name: ______________________ Date: __________

Multiplying in Raws

Directions: Solve the Multiplications

325 X 8 =

328 X 7 =

852 X 2 =

645 X 9 =

698 X 6 =

347 X 3 =

741 X 9 =

321 X 6 =

214 X 5 =

147 X 3 =

324 X 7 =

698 X 2 =

324 X 3 =

147 X 4 =

Name: ____________________ Date: __________

Multiplying in Raws

Directions: Solve the Multiplications

632 X 7 = ____________

358 X 4 = ____________

374 X 3 = ____________

987 X 4 = ____________

965 X 7 = ____________

741 X 7 = ____________

374 X 3 = ____________

584 X 3 = ____________

147 X 4 = ____________

369 X 8 = ____________

728 X 4 = ____________

628 X 6 = ____________

654 X 7 = ____________

415 X 9 = ____________

Name: ____________________ Date: __________

Multiplying in Raws

Directions: Solve the Multiplications

470 X 8 = ____________

870 X 7 = ____________

907 X 4 = ____________

908 X 7 = ____________

369 X 3 = ____________

471 X 7 = ____________

854 X 6 = ____________

698 X 8 = ____________

471 X 5 = ____________

754 X 3 = ____________

358 X 7 = ____________

418 X 7 = ____________

541 X 9 = ____________

354 X 5 = ____________

Solutions

Name: ______________________ Date: __________

1

Multiplying in Raws

Directions: Solve the Multiplications

60 X 8 = 480

87 X 7 = 609

90 X 9 = 810

85 X 9 = 765

65 X 7 = 455

66 X 8 = 528

36 X 9 = 324

70 X 8 = 560

99 X 3 = 297

60 X 7 = 420

64 X 8 = 512

65 X 2 = 130

98 X 7 = 686

30 X 8 = 240

Name: ______________________ Date: __________ 2

Multiplying in Raws

Directions: Solve the Multiplications

85 X 5 = 425

58 X 7 = 406

35 X 9 = 315

65 X 8 = 520

73 X 6 = 438

90 X 9 = 810

97 X 9 = 873

67 X 5 = 335

78 X 2 = 156

60 X 7 = 420

80 X 8 = 640

78 X 8 = 624

68 X 2 = 136

65 X 8 = 520

Name: ______________________ Date: __________

Multiplying in Raws

Directions: Solve the Multiplications

63 X 8 = 504

57 X 7 = 399

40 X 9 = 360

96 X 9 = 864

99 X 5 = 495

94 X 2 = 188

36 X 9 = 324

65 X 3 = 195

88 X 8 = 704

85 X 7 = 595

50 X 3 = 150

60 X 4 = 240

70 X 8 = 560

69 X 4 = 276

Name: ____________________ Date: __________

Multiplying in Raws

Directions: Solve the Multiplications

66 X 8 = 528

87 X 6 = 522

94 X 9 = 846

32 X 5 = 160

39 X 9 = 351

93 X 2 = 186

35 X 3 = 105

30 X 8 = 240

69 X 3 = 207

72 X 3 = 216

68 X 3 = 204

50 X 8 = 400

88 X 8 = 704

99 X 6 = 594

Name: ______________________ Date: __________

Multiplying in Raws

Directions: Solve the Multiplications

96 X 5 = 480

77 X 7 = 539

56 X 9 = 504

52 X 3 = 156

94 X 4 = 376

40 X 9 = 360

15 X 6 = 90

84 X 8 = 672

62 X 7 = 434

48 X 3 = 144

89 X 8 = 712

62 X 7 = 434

23 X 6 = 138

65 X 9 = 585

Name: ______________________ Date: __________

Multiplying in Raws

Directions: Solve the Multiplications

13 X 8 = 104

56 X 7 = 392

87 X 9 = 783

95 X 5 = 475

54 X 9 = 486

76 X 3 = 228

49 X 9 = 441

86 X 8 = 688

32 X 6 = 192

32 X 9 = 288

49 X 7 = 343

56 X 8 = 448

13 X 3 = 39

68 X 8 = 544

Name: ______________________________ Date: ____________

Multiplying in Raws

Directions: Solve the Multiplications

60 X 2 = 120

82 X 7 = 574

13 X 9 = 117

60 X 6 = 360

45 X 9 = 405

36 X 3 = 108

96 X 8 = 768

56 X 6 = 336

78 X 8 = 624

32 X 7 = 224

89 X 5 = 445

86 X 7 = 602

46 X 8 = 368

65 X 3 = 195

Name: ____________________ Date: __________

Multiplying in Raws

Directions: Solve the Multiplications

60 X 8 = 480

87 X 7 = 609

45 X 9 = 405

89 X 9 = 801

35 X 9 = 315

59 X 9 = 531

13 X 9 = 117

78 X 6 = 468

31 X 8 = 248

12 X 3 = 36

34 X 4 = 136

48 X 3 = 144

66 X 4 = 264

48 X 8 = 384

Name: ______________________ Date: ____________

Multiplying in Raws

Directions: Solve the Multiplications

65 X 8 = 520

87 X 3 = 261

56 X 9 = 504

17 X 3 = 51

94 X 9 = 846

79 X 7 = 553

12 X 6 = 72

60 X 4 = 240

95 X 6 = 570

75 X 3 = 225

15 X 9 = 135

61 X 7 = 427

56 X 4 = 224

42 X 7 = 294

Name: ______________________ Date: __________

10

Multiplying in Raws

Directions: Solve the Multiplications

70 X 8 = 560

78 X 7 = 546

32 X 8 = 256

19 X9 = 171

80 X 6 = 480

50 X 4 = 200

70 X 5 = 350

60 X 3 = 180

90 X 4 = 360

63 X 5 = 315

76 X 6 = 456

629 X 7 = 4403

63 X8 = 504

83 X 9 = 747

Name: ______________________ Date: __________

Multiplying in Raws

Directions: Solve the Multiplications

604 X 8 = 4832

874 X 7 = 6118

906 X 9 = 8154

806 X 3 = 2418

365 X 4 = 1460

635 X 6 = 3810

325 X 9 = 2925

160 X 7 = 1120

965 X 6 = 5790

341 X 4 = 1364

321 X 2 = 642

321 X 8 = 2568

874 X 4 = 3496

260 X 9 = 2340

Name: ______________________ Date: __________

Multiplying in Raws

Directions: Solve the Multiplications

632 X 8 = 5056

865 X 7 = 6055

954 X 9 = 8586

632 X 7 = 4424

852 X 3 = 2556

741 X 4 = 2964

852 X 7 = 5964

963 X 4 = 3852

369 X 6 = 2214

258 X 8 = 2064

852 X 9 = 7668

147 X 3 = 441

741 X 2 = 1482

123 X 1 = 123

Name: ______________________ Date: __________

Multiplying in Raws

Directions: Solve the Multiplications

632 X 8 = 5056

324 X 7 = 2268

987 X 9 = 8883

789 X 4 = 3156

654 X 9 = 5886

46 X 3 = 138

321 X 2 = 648

123 X 4 = 492

357 X 6 = 2142

753 X 7 = 5271

241 X 9 = 723

421 X 3 = 1263

689 X 3 = 2067

986 X 4 = 3944

Name: ____________________ Date: __________

Multiplying in Raws

Directions: Solve the Multiplications

160 X 8 = 1280

187 X 7 = 1309

930 X 6 = 5580

190 X 8 = 1520

290 X 9 = 2610

390 X 7 = 2730

901 X 4 = 3604

601 X 7 = 4207

609 X 6 = 3654

197 X 5 = 985

660 X 8 = 5280

670 X 3 = 2010

360 X 8 = 2880

960 X 3 = 2880

Name: ______________________ Date: ____________ 15

Multiplying in Raws

Directions: Solve the Multiplications

655 X 8 = 5240

834 X 7 = 5838

937 X 9 = 8433

347 X 4 = 1388

157 X 6 = 942

978 X 7 = 6846

670 X 9 = 6030

662 X 9 = 5958

675 X 7 = 4725

874 X 5 = 4370

324 X 3 = 972

478 X 1 = 478

324 X 9 = 2916

357 X 4 = 1428

Name: ____________________ Date: __________

Multiplying in Raws

Directions: Solve the Multiplications

160 X 4 = 640

835 X 7 = 5845

998 X 4 = 3992

955 X 9 = 8595

354 X 7 = 2478

741 X 3 = 2223

475 X 9 = 4275

632 X 8 = 5056

324 X 7 = 2268

345 X 8 = 2760

852 X 3 = 2556

320 X 4 = 1280

120 X 2 = 240

400 X 8 = 3200

Name: ______________________________ Date: __________ 17

Multiplying in Raws

Directions: Solve the Multiplications

460 X 8 = 3680

870 X 7 = 6090

900 X 6 = 5400

907 X 4 = 3628

905 X 9 = 8145

904 X 6 = 5424

903 X 9 = 8127

60 X 7 = 420

440 X 8 = 3520

604 X 9 = 5436

605 X 7 = 4235

607 X 6 = 3642

609 X 5 = 3045

608 X 4 = 2432

Name: ______________________ Date: __________

Multiplying in Raws

Directions: Solve the Multiplications

645 X 8 = 5160

745 X 7 = 5215

945 X 5 = 4725

907 X 4 = 3628

940 X 7 = 6580

908 X 9 = 8172

964 X 6 = 5784

605 X 5 = 3025

650 X 8 = 5200

604 X 4 = 2416

660 X 4 = 2640

606 X 9 = 5454

60 9 X 6 = 3654

960 X 5 = 4800

Name: ______________________ Date: ____________

Multiplying in Raws

Directions: Solve the Multiplications

640 X 4 = 2560

699 X 4 = 2796

874 X 3 = 2622

542 X 3 = 1626

905 X 9 = 8145

789 X 9 = 7101

940 X 7 = 6580

657 X 6 = 3942

490 X 8 = 3920

325 X 7 = 2275

690 X 7 = 4830

624 X 1 = 624

966 X 6 = 5796

723 X 7 = 5061

Name: ______________________ Date: __________

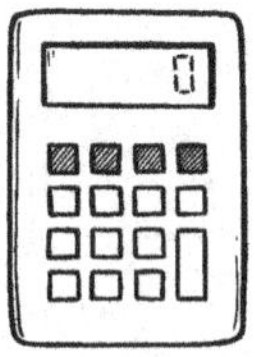

Multiplying in Raws

Directions: Solve the Multiplications

524 X 3 = 1572

324 X 4 = 1296

354 X 5 = 1770

852 X 9 = 7668

754 X 7 = 5278

957 X 8 = 7656

895 X 4 = 3580

507 X 6 = 3042

602 X 8 = 4816

666 X 8 = 5328

555 X 7 = 3885

688 X 8 = 5504

428 X 9 = 3852

625 X 4 = 2500

Name: ____________________ Date: __________

Multiplying in Raws

Directions: Solve the Multiplications

602 X 8 = 4816

870 X 7 = 6090

904 X 7 = 6328

940 X 6 = 5640

999 X 8 = 7992

690 X 5 = 3450

490 X 9 = 4410

460 X 8 = 3680

640 X 4 = 2560

605 X 6 = 3630

470 X 8 = 3760

604 X 7 = 4228

350 X 5 = 1750

352 X 8 = 2816

Name: ______________________________ Date: __________

Multiplying in Raws

Directions: Solve the Multiplications

622 X 8 = 4976

487 X 7 = 3409

241 X 3 = 723

123 X 9 = 1107

321 X 2 = 642

742 X 1 = 742

324 X 4 = 1296

412 X 4 = 1648

644 X 2 = 1288

365 X 6 = 2190

321 X 8 = 2568

654 X 9 = 5886

321 X 4 = 1284

560 X 6 = 3360

Name: ______________________________ Date: ____________

Multiplying in Raws

Directions: Solve the Multiplications

222 X 8 = 1776

833 X 2 = 1666

753 X 4 = 3012

415 X 3 = 1245

652 X 5 = 3260

872 X 2 = 1744

358 X 4 = 1432

325 X 4 = 1300

685 X 4 = 2740

389 X 2 = 778

952 X 2 = 1904

620 X 4 = 2480

333 X 4 = 1332

678 X 6 = 4068

Name: ______________________________ Date: __________

Multiplying in Raws

Directions: Solve the Multiplications

322 X 8 = 2576

332 X 7 = 2324

666 X 9 = 5994

878 X 9 = 7902

989 X 9 = 8901

323 X 9 = 2907

454 X 9 = 4086

545 X 8 = 4360

656 X 8 = 5248

678 X 8 = 5424

878 X 8 = 7024

654 X 8 = 5232

678 X 8 = 5424

603 X 8 = 4824

Name: ______________________ Date: __________

Multiplying in Raws

Directions: Solve the Multiplications

635 X 8 = 5080

854 X 7 = 5978

546 X 9 = 4914

656 X 9 = 5904

780 X 9 = 7020

879 X 9 = 7911

935 X 9 = 8415

451 X 8 = 3608

612 X 8 = 4896

678 X 8 = 5424

541 X 8 = 4328

645 X 8 = 5160

785 X 8 = 6280

784 X 8 = 6272

Name: ______________________ Date: __________

Multiplying in Raws

Directions: Solve the Multiplications

124 X 8 = 992

410 X 3 = 1230

325 X 7 = 2275

147 X 4 = 588

142 X 2 = 284

159 X 6 = 954

921 X 9 = 8289

241 X 8 = 1928

121 X 4 = 484

231 X 7 = 1617

541 X 7 = 3787

312 X 4 = 1248

213 X 5 = 1065

412 X 9 = 3708

Name: ____________________ Date: ____________

Multiplying in Raws

Directions: Solve the Multiplications

325 X 4 = 1300

210 X 6 = 1260

365 X 4 = 1460

782 X 7 = 5474

310 X 1 = 310

120 X 7 = 840

310 X 8 = 2480

412 X 7 = 2884

321 X 8 = 2568

410 X 9 = 3690

321 X 3 = 963

632 X 4 = 2528

310 X 5 = 1550

103 X 6 = 618

Name: ______________________ Date: __________

Multiplying in Raws

Directions: Solve the Multiplications

363 X 2 = 726

333 X 6 = 1998

444 X 3 = 1332

454 X 4 = 1816

654 X 7 = 4578

231 X 5 = 1155

621 X 9 = 5589

632 X 8 = 5056

665 X 4 = 2660

641 X 4 = 2564

324 X 6 = 1944

753 X 4 = 3012

635 X 7 = 4445

852 X 8 = 6816

Name: ______________________ Date: __________

Multiplying in Raws

Directions: Solve the Multiplications

258 X 8 = 2064

369 X 7 = 2583

365 X 9 = 3285

357 X 4 = 1428

753 X 6 = 4518

951 X 5 = 4755

158 X 3 = 474

336 X 8 = 2688

584 X 9 = 5256

678 X 4 = 2712

324 X 8 = 2592

740 X 6 = 4440

625 X 3 = 1875

471 X 7 = 3297

Name: ______________________ Date: __________

Multiplying in Raws

Directions: Solve the Multiplications

632 X 8 = 5056

545 X 7 = 3815

941 X 9 = 8469

321 X 4 = 1284

784 X 4 = 3136

197 X 3 = 591

379 X 7 = 2653

147 X 7 = 1029

247 X 5 = 1235

374 X 4 = 1500

314 X 8 = 2512

413 X 6 = 2478

521 X 9 = 4689

332 X 5 = 1660

Name: ______________________ Date: __________

Multiplying in Raws

Directions: Solve the Multiplications

475 X 5 = 2375

425 X 7 = 2975

214 X 6 = 1284

354 X 8 = 2832

923 X 7 = 6461

123 X 4 = 492

143 X 3 = 429

234 X 6 = 1404

741 X 7 = 5187

241 X 5 = 1205

149 X 3 = 447

324 X 3 = 972

324 X 9 = 2916

324 X 8 = 2592

Name: ______________________ Date: ____________

Multiplying in Raws

Directions: Solve the Multiplications

241 X 7 = 1687

147 X 7 = 1029

598 X 9 = 5382

398 X 4 = 1592

385 X 3 = 1155

347 X 2 = 694

541 X 1 = 541

341 X 3 = 1023

341 X 4 = 1364

324 X 8 = 1296

314 X 4 = 1256

214 X 9 = 1926

412 X 7 = 2884

214 X 6 = 1284

Name: ______________________ Date: __________

Multiplying in Raws

Directions: Solve the Multiplications

632 X 8 = 5056

870 X 7 = 6090

930 X 1 = 930

901 X 6 = 5406

321 X 4 = 1284

214 X 4 = 856

321 X 6 = 1926

604 X 5 = 3020

641 X 3 = 1923

147 X 8 = 1176

324 X 5 = 1620

147 X 8 = 1176

147 X 6 = 882

247 X 4 = 988

Name: ______________________________ Date: ____________

Multiplying in Raws

Directions: Solve the Multiplications

148 X 3 = 444

741 X 7 = 5187

564 X 2 = 1128

247 X 4 = 988

324 X 9 = 2916

214 X 3 = 642

327 X 5 = 1635

214 X 9 = 1926

325 X 7 = 2275

574 X 4 = 2296

324 X 8 = 2592

329 X 2 = 658

951 X 1 = 951

217 X 7 = 1519

Name: ______________________ Date: __________

Multiplying in Raws

Directions: Solve the Multiplications

247 X 9 = 2223

231 X 7 = 1617

654 X 2 = 1308

456 X 3 = 1368

358 X 3 = 1074

184 X 9 = 1656

247 X 7 = 1729

284 X 6 = 1704

398 X 8 = 3184

574 X 7 = 4018

147 X 6 = 882

214 X 3 = 642

320 X 7 = 2240

260 X 8 = 2080

Name: ______________________ Date: __________

Multiplying in Raws

Directions: Solve the Multiplications

332 X 8 = 2656

854 X 4 = 3416

324 X 9 = 2916

369 X 2 = 738

150 X 3 = 450

724 X 8 = 5792

231 X 7 = 1617

324 X 6 = 1944

624 X 3 = 1872

457 X 7 = 3199

325 X 4 = 1300

124 X 2 = 248

214 X 7 = 1498

124 X 8 = 992

Name: ______________________ Date: __________

Multiplying in Raws

Directions: Solve the Multiplications

247 X 4 = 988

321 X 7 = 2247

254 X 4 = 1016

369 X 9 = 3321

914 X 4 = 3656

321 X 6 = 1926

358 X 7 = 2506

258 X 8 = 2064

247 X 9 = 2223

241 X 7 = 1687

147 X 8 = 1176

214 X 6 = 1284

147 X 8 = 1176

324 X 7 = 2268

Name: ______________________ Date: __________

Multiplying in Raws

Directions: Solve the Multiplications

584 X 5 = 2920

325 X 7 = 2275

147 X 4 = 588

258 X 4 = 1032

214 X 7 = 1498

179 X 8 = 1432

721 X 9 = 6489

187 X 7 = 1309

641 X 8 = 5128

274 X 9 = 2466

198 X 7 = 1386

384 X 8 = 3072

444 X 9 = 3996

147 X 3 = 441

Name: ______________________ Date: ____________

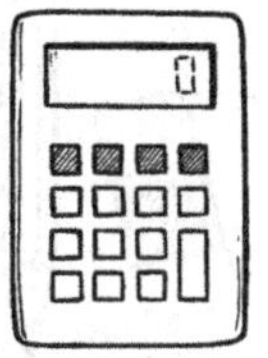

Multiplying in Raws

Directions: Solve the Multiplications

641 X 8 = 5128

120 X 7 = 840

324 X 4 = 1296

147 X 9 = 1323

258 X 2 = 516

247 X 3 = 741

932 X 8 = 7456

471 X 5 = 2355

666 X 6 = 3996

250 X 7 = 1750

214 X 4 = 856

324 X 2 = 648

147 X 3 = 441

324 X 5 = 1620

Name: ______________________ Date: ____________

Multiplying in Raws

Directions: Solve the Multiplications

741 X 7 = 5187

145 X 9 = 1305

241 X 6 = 1446

369 X 7 = 2583

147 X 5 = 735

347 X 8 = 2776

234 X 7 = 1638

257 X 7 = 1799

247 X 4 = 988

147 X 2 = 294

324 X 5 = 1620

741 X 8 = 5928

235 X 6 = 1410

124 X 4 = 496

Name: ____________________ Date: __________

Multiplying in Raws

Directions: Solve the Multiplications

123 X 7 = 861

124 X 4 = 496

124 X 3 = 372

912 X 2 = 1824

124 X 9 = 1116

324 X 7 = 2268

147 X 4 = 588

147 X 8 = 1176

369 X 6 = 2214

147 X 7 = 1029

478 X 5 = 2390

124 X 9 = 1116

147 X 8 = 1176

475 X 4 = 1900

Name: ______________________ Date: __________

Multiplying in Raws

Directions: Solve the Multiplications

332 X 4 = 1328

325 X 6 = 1950

541 X 9 = 4869

241 X 8 = 1928

327 X 7 = 2289

147 X 5 = 735

954 X 9 = 8586

471 X 7 = 3297

147 X 8 = 1176

147 X 4 = 588

150 X 6 = 900

328 X 8 = 2624

147 X 9 = 1323

149 X 6 = 894

Name: ______________________ Date: __________

Multiplying in Raws

Directions: Solve the Multiplications

324 X 8 = 2592

257 X 7 = 1799

147 X 4 = 588

324 X 6 = 1944

247 X 9 = 2223

954 X 7 = 6678

894 X 3 = 2682

364 X 7 = 2548

968 X 8 = 7744

784 X 7 = 5488

324 X 9 = 2916

784 X 6 = 4704

214 X 4 = 856

320 X 3 = 960

Name: ______________________________ Date: __________

Multiplying in Raws

Directions: Solve the Multiplications

623 X 8 = 4984

870 X 7 = 6090

901 X 9 = 8109

999 X 7 = 6993

331 X 6 = 1986

147 X 5 = 735

324 X 3 = 972

147 X 3 = 441

789 X 8 = 6312

324 X 7 = 2268

985 X 3 = 2955

394 X 7 = 2758

459 X 8 = 3672

395 X 3 = 1185

Name: ____________________ Date: __________

Multiplying in Raws

Directions: Solve the Multiplications

758 X 8 = 6064

954 X 3 = 2862

347 X 7 = 2429

366 X 9 = 3294

247 X 7 = 1729

124 X 2 = 248

369 X 8 = 2952

827 X 8 = 6616

365 X 7 = 2555

369 X 4 = 1476

741 X 2 = 1482

987 X 8 = 7896

475 X 4 = 1900

854 X 8 = 6832

Name: ______________________ Date: __________

Multiplying in Raws

Directions: Solve the Multiplications

374 X 3 = 1122

942 X 2 = 1884

248 X 2 = 496

741 X 3 = 2223

240 X 9 = 2160

254 X 7 = 1778

371 X 8 = 2968

124 X 8 = 992

731 X 6 = 4386

412 X 6 = 2472

914 X 4 = 3656

214 X 8 = 1712

419 X 9 = 3771

924 X 2 = 1848

Name: ______________________ Date: __________

Multiplying in Raws

Directions: Solve the Multiplications

324 X 8 = 2592

214 X 9 = 1926

124 X 4 = 496

987 X 2 = 1974

214 X 1 = 214

941 X 6 = 5646

341 X 3 = 1023

241 X 8 = 1928

365 X 7 = 2555

941 X 8 = 7528

241 X 9 = 2169

874 X 3 = 2622

954 X 2 = 1908

632 X 8 = 5056

Name: ______________________ Date: ____________

Multiplying in Raws

Directions: Solve the Multiplications

325 X 8 = 2600

328 X 7 = 2296

852 X 2 = 1704

645 X 9 = 5805

698 X 6 = 4188

347 X 3 = 1041

741 X 9 = 6669

321 X 6 = 1926

214 X 5 = 1070

147 X 3 = 441

324 X 7 = 2268

698 X 2 = 1396

324 X 3 = 972

147 X 4 = 588

Name: ____________________ Date: ____________

Multiplying in Raws

Directions: Solve the Multiplications

632 X 7 = 4424

358 X 4 = 1432

374 X 3 = 1122

987 X 4 = 3948

965 X 7 = 6755

741 X 7 = 5187

374 X 3 = 1122

584 X 3 = 1752

147 X 4 = 1029

369 X 8 = 2952

728 X 4 = 2912

628 X 6 = 3768

654 X 7 = 4578

415 X 9 = 3735

Name: ______________________ Date: ____________

Multiplying in Raws

Directions: Solve the Multiplications

470 X 8 = 3760

870 X 7 = 6090

907 X 4 = 3628

908 X 7 = 6356

369 X 3 = 1107

471 X 7 = 3297

854 X 6 = 5124

698 X 8 = 5584

471 X 5 = 2355

754 X 3 = 2262

358 X 7 = 2506

418 X 7 = 2926

541 X 9 = 4869

354 X 5 = 1770

Made in the USA
Monee, IL
02 November 2026

67383641R00057